AGSO Phanerozoic Timescale 1995

Wallchart and Explanatory Notes

P.J. Jones (compiler) with personnel of the AGSO Timescale Calibration and Development Project 1995.

A new Phanerozoic chart and brief overview of the history and current status of its major subdivisions has been compiled, based on detailed charts and reviews presented in a larger reference work, *An Australian Phanerozoic Timescale* [*APT*] (Young & Laurie, in press). This more comprehensive work, itself a compilation and re-evaluation of previously published data, is an attempt to calibrate isotopic age and biochronologic data to determine reliable tie-points, and an attempt to interpolate the calibration of the intervals between them. The methodology used for its construction is selective, and confined to those biochronologic and isotopic data considered reliable.

Ages adopted for major boundaries are: Precambrian–Cambrian [545 Ma], Cambrian–Ordovician [490 Ma, or younger], Ordovician–Silurian [434 Ma], Silurian–Devonian [410 Ma], Devonian–Carboniferous [354 Ma], Carboniferous–Permian [298 Ma], Permian–Triassic [251 Ma], Triassic–Jurassic [205 Ma], Jurassic–Cretaceous [141 Ma], Cretaceous–Palaeogene [65 Ma], Palaeogene/Neogene [23.8 Ma], Tertiary–Quaternary [1.78 Ma].

APT places emphasis on zonal schemes derived from Australian biostratigraphy; it integrates standard European time units and Australian Cambrian and Ordovician stages. Initially prepared for use in AGSO and its STRATDAT database, this scale is significant for global correlation of Phanerozoic sequences and their biotas. Interpretation of timescale calibration will continue to be revised through the integration of new multidisciplinary research (biochronology, geochronology, magnetostratigraphy). As significant new information becomes available, revised versions of the timescale will be published.

Introduction

On a global basis, biostratigraphy provides the direction of time in the strata, but cannot, of itself, measure durations. It provides a relative scale that, based on stratigraphic sequences with subdivisions defined in reference sections (boundary stratotypes), has become known as the global stratigraphic scale (e.g. the **chronostratic** scale of Harland et al. 1990). A geologic timescale is the union of this relative global stratigraphic scale with a numerical scale based on units of duration, that is, a scale of estimated age (e.g. the **chronometric** scale of Harland et al. 1990).

The numerical calibration of the global stratigraphic scale is provided by isotopic geochronology, based on the known rates of decay of radioactive elements. Advances in the use of this principle have been extensive, with application to an ever-widening range of decay schemes and minerals. Early geochronology relied on dating intrusive rocks, but now reliable ages are measured for volcanic rocks that have a more direct relationship with stratigraphy. In parallel with these developments, there have been recent major advances in the technology of dating, for example the Sensitive High Resolution Ion MicroProbe (SHRIMP), built at the Australian National University, and developed as a joint venture with the Australian Geological Survey Organisation (AGSO), and the development of laser $^{40}Ar/^{39}Ar$ dating.

Implicit in the construction of a geologic timescale is the question of inter-laboratory accuracy in dating. Age measurements are never 'absolute', but are always relative to a value ascribed to a laboratory standard. Measurements from one laboratory cannot be validly compared with those from another, unless all data are referred to an 'agreed' standard. The estimation of boundary dates is by interpolation because isotopically dated samples are seldom associated directly with a particular biozone or boundary. Instead, the age of the biozone or boundary must be interpolated from the measured ages of the nearest dated volcanic horizons above and below. That interpolation is based on correlation with some attribute of the stratigraphic scale, such as number of biozones, or number of magnetic intervals. It follows that the calibration of the numerical, biostratigraphic and other scales is improved by iteration, an essential part of the method, and the product is an *estimated age scale.*

Recent years have seen a rapid growth in other methods of correlation of global geological events, such as reversals in the Earth's magnetic field (magnetostratigraphy), and secular variation in element and isotopic compositions such as C, O and Sr (chemostratigraphy). Each method needs to be calibrated with the estimated age scale provided by geochronology, and all should be integrated to form a reliable geologic timescale.

This publication is a summary of a more expansive work, *An Australian Phanerozoic Timescale* (Young & Laurie, in press). Included is a calibrated geochronologic chart, or timescale, and a brief overview of the history and current status of major subdivisions for each of the geological periods. The IUGS-ratified *Precambrian Timescale* (Plumb 1991) is also included. A linear chronometric timescale has been applied to a Phanerozoic chart taken down to the level of stages. The calibrations presented here are subject to future refinement in the light of new data or other modes of interpolation. Therefore we fully anticipate changes in the geologic timescale we have compiled, and distinguish the present edition by incorporating the year of publication within its title.

An Australian Phanerozoic Timescale

An Australian Phanerozoic Timescale (*APT*; Young & Laurie, in press) originated in the Bureau of Mineral Resources with preparation of biostratigraphic charts for the *Palaeogeographic Atlas of Australia* project and its successor, the *Phanerozoic History of Australia* project. Both projects were sponsored by the Australian Petroleum Industry Research Association (APIRA). Revised and updated versions of these charts, together with explanatory notes, were published in 1989–91 as a series of ten *BMR Records* (1989/31–40). The purpose was to provide a set of calibrated biostratigraphic charts specifically for use in the Australian region, in recognition of the fact that detailed local age control is a prerequisite to interpreting the development and structure of the sedimentary basins that host major petroleum, coal, and sedimentary mineral deposits.

Several recently published geological timescales (e.g. Odin 1982, 1994; Snelling 1985; Cowie & Bassett 1989; Harland et al. 1990; Odin & Odin 1990) summarise the current ideas on internationally agreed subdivisions of geological time. Many of these international stages, of northern hemisphere origin, contain cosmopolitan biotas, and biozonations that are directly applicable to Australian sedimentary basins, for example the Late Devonian conodont and ammonoid zonations of the Canning Basin (Glenister & Klapper 1966; Becker et al. 1993). However, for some intervals of the Phanerozoic, Australia had distinctive endemic (Gondwanan) biotas, and much of the detailed biocorrelation within and between Australian basins is therefore based on zonal schemes derived from local biostratigraphic studies. Thus, the immediate objective of *APT* is the integration of Australian biozonal information into a detailed timescale in order to satisfy the need of modern basin analysis for high-resolution age control.

APT, like other timescales, is a compilation and re-evaluation of previously published data, an attempt to calibrate isotopic age and biochronologic data to determine reliable tie-points, and an attempt to interpolate the calibration of the intervals between them. It differs from most international timescales in its emphasis on zonal schemes derived from Australian biostratigraphy, and its integration of standard European time units and Australian Cambrian and Ordovician stages. The methodology does not compare with that of the computer-fit scales of Harland et al. (1990) and Gradstein et al. (1994); instead, a selective use has been made of those biochronologic and isotopic age data considered reliable; in this we follow Odin (1982, 1994) and Snelling (1985). Initially prepared for in-house use — for example, *APT* underpins the AGSO STRATDAT database — this scale is significant for global correlation of Phanerozoic sequences and their biotas.

Interpretation of timescale calibration will continue to be revised through the integration of new multidisciplinary research (biochronology, geochronology, magnetostratigraphy). As significant new information becomes available, revised versions of the timescale will be published.

Precambrian

The Precambrian, also called the Cryptozoic — the period of 'hidden life' (Chadwick 1930) — refers to the approximately 4 billion years [4 Ga] preceding the Phanerozoic. Uncertainty of detailed stratigraphic correlation and the relative recency of detailed Precambrian stratigraphic research result in subdivisions that are not as detailed as those of the Phanerozoic — an imbalance being redressed now through geochronometry, a developing biostratigraphy, and the more recent applications of sequence stratigraphy, and magnetostratigraphy.

The Subcommission of Precambrian Stratigraphy (SPS) has recently advanced a strictly chronometric division of the Precambrian Timescale, and this has been ratified by the International Union of Geological Sciences (IUGS) (Plumb 1991). Precambrian time is divided into two eons, the Archaean and Proterozoic, which together represent about 88% of the Earth's history.

Archaean Eon

The Archaean Eon, as a standard time division, is taken to end at 2.5 Ga. As this term originally referred to the oldest known rocks, no lower boundary was defined. Harland et al. (1990) arbitrarily selected an age of 4.0 Ga for the base, but rocks as old as 4.0 Ga are preserved in North America, while detrital zircons found in Western Australian metasediments (dated at 4.3–4.2 Ga; Froude et al. 1983; Compston & Pidgeon 1986) are evidence that buoyant crust existed on Earth prior to 4.2 Ga. Although some regional schemes for the subdivision of Archaean time have been proposed (see Harland et al. 1990) an internationally acceptable subdivision has yet to be ratified. The oldest Archaean stromatolites have been found in well-preserved sedimentary rocks dated at 3.2 Ga (Lowe 1994). During the rest of the Archaean there was slow diversification of procaryotes (single-celled organisms lacking cell nuclei).

Proterozoic Eon

The Proterozoic Eon (2.5–0.545 Ga) is divided into three eras: Palaeoproterozoic (2500–1600 Ma), Mesoproterozoic (1600–1000 Ma), and Neoproterozoic (1000–545 Ma). Each era is also divided chronometrically into periods, named after 'globally significant geological events'.

The Palaeoproterozoic Era is divided into four periods: Siderian (2500–2300 Ma), Rhyacian (2300–2050 Ma), Orosirian (2050–1800 Ma), and Statherian (1800–1600 Ma).

The Mesoproterozoic Era is divided into three periods: Calymmian (1600–1400 Ma), Ectasian (1400–1200 Ma) and Stenian (1200–1000 Ma).

Discussion of these periods lies outside the scope of this summary. However, a brief discussion of the Neoproterozoic Era, and especially its youngest period, is given because of its pertinence in defining the beginning of the Palaeozoic Era and hence the Phanerozoic Eon.

NEOPROTEROZOIC ERA (1000–545 Ma)

The chronometric definition of the Neoproterozoic Era (Plumb 1991) can be broadly aligned with the Adelaidean 'System/Period' as stratigraphically defined in the Adelaide Fold Belt (Preiss 1987). The Adelaidean has a duration comparable with that of the Palaeozoic, and perhaps therefore should be regarded as an era. The Neoproterozoic Era is divided into three periods: Tonian (1000–850 Ma), Cryogenian (850–?650 Ma) and an informal 'Neoproterozoic III' (from ?650 Ma to the base of Cambrian). The late Cryogenian may correspond to the Sturtian (glacial) Period, but present geochronologic data are insufficient to correlate the SPS chronometric scale with the Adelaidean sequence of South Australia with any confidence.

'Neoproterozoic III' Period (650?-545 Ma) At present this period is an informal unit, recommended by the SPS as an interim measure because an international decision has yet to be made regarding its formal name (after a global stratigraphic standard) and precise chronometric limits (Plumb 1991). The lower limit of 'Neoproterozoic III' has been given a provisional age of 650 Ma, which is subject to the age of the boundary stratotype for the latest Precambrian global stratigraphic unit, yet to be selected (Plumb 1991). The names of two global stratigraphic units (Ediacaran and Vendian) are available for the replacement of the interim 'Neoproterozoic III'. The Ediacaran Period, based on sequences defined in Australia, is not officially recognised by the SPS.

Ediacaran Period (610?–545 Ma) The base of the Ediacaran is placed at the base of the Wonoka Formation (in the post-glacial uppermost part of the Adelaidean) in Bunyeroo Gorge in the northern Flinders Ranges, South Australia (Jenkins 1981, 1990).

The Ediacaran Period contains the remains, impressions and traces of the earliest possible soft-bodied metazoans, thought to represent hydrozoans, scyphozoans, medusoids, cnidarians, annelids, and arthropods (Glaessner 1984). Although many workers support such assignments to metazoan groups, an alternative hypothesis refers to them as a distinct group, the Vendobionta (Seilacher 1992), with no known Phanerozoic counterpart. This Ediacaran fauna has a relatively short history, and most of its taxa were extinct before the Cambrian (Campbell 1990). It is also the time of the first appearance of metaphytes (Zhang 1989). The Ediacaran fauna has a wide global distribution ranging from Australia to China, European and Asiatic Russia, England, Namibia, Newfoundland, eastern USA, and north-western Canada. The late Ediacaran fauna of the stratotype can be correlated with a similar fauna in the Valdai Group of the Vendian *sensu stricto* of northern Russia (Keller 1979). On the basis of K/Ar glauconite dating, Keller (1979) considers the age of the base of the Valdai Group to be about 620–600 Ma. In a review of the Precambrian–Cambrian boundary in south-eastern Newfoundland and adjacent areas (Avalon Zone), Conway Morris (1989) showed that the age of the earliest Ediacaran biota is constrained by the age of a volcanic/plutonic episode (Harbour Main Group/Holyrood Granite) at about 590–620 Ma. Thus the base of the Ediacaran Period is provisionally placed on our chart at 610 Ma.

An alternative concept of the Ediacaran (the **Ediacaran** Period of Cloud & Glaessner 1982) places the lower boundary at a lower level in the stratotype: at the base of the first stratigraphic unit (Nuccaleena Formation), which occurs above the highest tillite-containing deposits (Elatina Formation), and prior to the appearance of known Ediacaran fossils. This level is at the base of the Wilpena Group, and therefore the concept includes the post-glacial phase of 'Neoproterozoic III', for which an acritarch biostratigraphy has recently been proposed (Grey 1994). It was chosen on the basis of the occurrence of a possible medusoid impression in the lower part of the Wilpena Group (Cloud & Glaessner 1982).

A review of both concepts has been made by Preiss (1987), who prefers to use the name Ediacaran, but in the wider sense of Cloud & Glaessner (1982). It has yet to be decided internationally whether: (i) the Ediacaran *sensu* Jenkins should rank as a Period on a par with the Cambrian, or as an epoch (series) of a Vendian Period (Harland et al. 1990), and (ii) the Ediacaran *sensu* Cloud & Glaessner should have priority over the Vendian Period.

Phanerozoic Eon (545 – 0 Ma)

The Phanerozoic Eon represents about 12% of the Earth's history and is divided into three eras: Palaeozoic (545–251 Ma), Mesozoic (251–65 Ma), and Cainozoic (65–0 Ma).

In Australia, the presence of sedimentary rocks belonging to each of the ten geological periods that constitute the Phanerozoic Eon was first established only when appropriate fossils were discovered. Preliminary palaeontological investigations of Australian sedimentary basins were used during the nineteenth century to establish the relative age of major suites of stratified rocks, and this provided a framework for more detailed subsequent biostratigraphic research. For example, graptolites were used to subdivide Ordovician strata in the Victorian goldfields at the beginning of this century. The Victorian sequence of graptolite zones is now one of the most finely subdivided in the world and is used as an international standard for the Ordovician Period (Cooper & Lindholm 1990).

Today the biostratigraphic control of Australian sedimentary sequences is based on a wide range of organisms, from microscopic remains such as unicellar diatoms, foraminiferids, and the pollen and spores of plants, to macroscopic and megascopic forms such as the larger invertebrates, and vertebrate remains ranging from fishes to mammals. Many groups widely used in the northern hemisphere are well represented in Australia, but have not yet received the detailed research needed for their full utilisation in biochronology.

In the accompanying chart Australian stages and names are shown for the Middle and Late Cambrian, and the entire Ordovician. For most other Phanerozoic periods, Australian biozones can be correlated with those of the standard international stratigraphic scales.

Palaeozoic Era (545–251 Ma)

CAMBRIAN PERIOD (545–490 Ma)

The Cambrian System was first recognised and named by Adam Sedgwick (in Sedgwick & Murchison 1835) for exposures in northern Wales, the name being derived from the Roman name for the region, *Cambria*. Modern international usage restricts the name essentially to Sedgwick's (1852) revised concept of a Lower Cambrian series (see also the discussion on the Ordovician and Silurian). The Cambrian Period worldwide is marked by the first abundant occurrence of mineralised body fossils of multicellular animals (the Metazoa). Some groups, such as the trilobites, archaeocyathids, and brachiopods, occur in great diversity, and are widely distributed in rocks of this age throughout the world.

The base of the Cambrian is defined in the global stratotype section and point (GSSP) recently ratified by the IUGS, on the Burin Peninsula of eastern Newfoundland (Landing 1994). In that section the Precambrian–Cambrian boundary is defined at the base of the ichnofossil zone of *Phycodes pedum*. The section contains three globally correlatable trace fossil zones occurring slightly earlier than the first trilobite faunas.

New zircon U/Pb age measurements in biostratigraphically well established horizons have brought about a radical recalibration of biozones to the Cambrian timescale. The Cambrian stages shown on the accompanying chart are scaled by interpolation against this new, but still sparse, database of zircon ages. The base of the Cambrian is estimated at 545 Ma, close to the age of 543.9$\pm$0.24 Ma, from conventional zircon U/Pb dating, for a volcanic breccia near the base of the *Anabarites trisulcata* Zone, in the earliest Nemakit-Daldynian (Manykaian) of northern Yakutia (Bowring et al. 1993). A U/Pb zircon age of 530.7$\pm$0.9 Ma for a supposed late 'Manykaian' volcanic ash from the Avalon terrane in New Brunswick, eastern Canada (Isachsen et al. 1994) suggests that the Nemakit-Daldynian–Tommotian boundary may be slightly younger than the 534 Ma shown on our chart. No data are available for the age of the Early–Middle Cambrian boundary, but an estimate of 509 Ma is used herein. The Mount Read Volcanics of Tasmania (Perkins & Walshe 1993), which contain Undillan to Boomerangian (late Middle Cambrian) trilobites, have SHRIMP zircon ages near 503 Ma, which suggest that the Middle–Late Cambrian boundary is younger than 500 Ma.

None of the epoch names available for the Cambrian Period elsewhere are currently applied to the Cambrian of Australia. Instead, the tripartite division of Early, Middle and Late Cambrian provides an informal basis, in which thirteen stages are recognised (Shergold 1989). The Early Cambrian is best developed in the Adelaide Fold Belt of South Australia (Zhuravlev & Gravestock 1994) and, by using archaeocyathid biostratigraphy, its stadial subdivision is correlated with that of Siberia. The Middle Cambrian stages are Australian, being extensive refinements by Öpik (1979) of agnostid zonations originally developed in Scandinavia. The Late Cambrian stages, entirely Australian, are based mainly on trilobite faunas (Öpik 1963, 1967; Shergold 1975, 1993), the exception being the uppermost stage (Datsonian), which is based on conodonts (Jones et al. 1971).

ORDOVICIAN PERIOD (490–434 Ma)

Charles Lapworth (1879) introduced the 'Ordovician' as a new system, to resolve the controversy over strata in the north Wales succession, which Sedgwick had included as Cambrian, but which Murchison regarded as belonging to his Silurian System (see below). In Britain acceptance of the new name was slow, and one of the earliest usages of the term 'Ordovician' was by the Australian-born and trained palaeontologist T.S. Hall (1897), in the first of a series of classic papers on graptolites from the Victorian goldfields.

The approximate level for the base of the Ordovician in the type area of northern Wales (the base of the Tremadoc Series) is represented by a hiatus, and a more complete Cambrian–Ordovician boundary stratotype section is currently being sought. Pelagic trilobites and graptolites have a long history of usage in Ordovician biostratigraphy, but recently conodont zonation has assumed greater importance. Currently an internationally correlatable conodont– graptolite datum is being considered at a section in north-eastern China. This datum lies close to the base of the *Cordylodus lindstromi* conodont zone, which is recognised in Australian sequences, and is used here as a provisional index fossil for defining the base of the Ordovician (see Webby & Nicoll 1989; Nicoll 1990, 1991; Shergold & Nicoll 1992).

The Ordovician was originally founded on a twofold division in Britain (Whittington & Williams 1964); however, a threefold division has been the common worldwide practice, albeit with little or no consistency between the divisions (Jaanusson 1960). The major subdivisions are into Early Ordovician (Tremadoc, Arenig, and Llanvirn series) and the Late Ordovician (Caradoc and Ashgill series). The Llandeilo, one of the six classical series as recognised in the type area, is now mostly incorporated within the base of the Caradoc Series, as modified by Fortey *et al.* (1995), and the Llanvirnian is extended to include the lowermost part of the classical Llandeilo, as a stage — the Llandeilian (*sensu* Elles 1925). In Australia the same twofold division of the

Ordovician is recognised, and is subdivided into nine stages based on the deepwater graptolitic succession of Victoria (VandenBerg & Cooper 1992). No equivalent stratigraphic standard scale exists for the shelly platform sequences of central Australia, except for the Warendan Stage, the lowermost stage based on conodonts.

No reliable isotope dates are available for the base of the Ordovician. Accepting a Middle–Late Cambrian boundary age of less than 500 Ma (Perkins & Walshe 1993), and allowing for the accumulation of Upper Cambrian strata, it must be considerably younger than 500 Ma, and is here arbitrarily placed at 490 Ma. The top of the Arenig, near 465 Ma, is constrained by a SHRIMP zircon age of 462±3 Ma within the Llanvirn Llanrin Volcanics (Compston & Williams 1992). Conventional measurement of zircons from the same sample gives an age of 465.7±2.1 Ma and would require an older top to the Arenig (Tucker et al. 1990). Reliable dates within the Ordovician, such as the Millbrig bentonite of eastern USA (Huff & Kolata 1990), accurately establish an age control of the middle part of the Ordovician. Higher parts of the Ordovician are scaled by a single SHRIMP age determination of 448±4 Ma for the mid-Caradoc (Longvillian Stage) Pont-y-ceunant Ash (Compston & Williams 1992). Conventional measurement for the same mineral separate analysed by Tucker et al. (1990) gives an age of 457.2±2.2 Ma.

SILURIAN PERIOD (434–410 Ma)

The Silurian Period was established in 1839 by R.I. Murchison to include a faunally distinct series of Welsh rocks included in the Cambrian by Sedgwick; the name derives from the Welsh Borderland tribe known to the Romans as *Silurii.* Lapworth (1879) subsequently incorporated Murchison's 'Lower Silurian' in his Ordovician System, and restricted Murchison's 'Upper Silurian' to approximately what are now the accepted limits of the Silurian System. Despite widespread usage, both Ordovician and Silurian systems only received formal international recognition at the 1960 IGC. In the ensuing 30 years all subdivisions (series and stages) of the Silurian have been internationally agreed and formally defined. The major subdivisions are the Early Silurian (Llandovery and Wenlock series) and the Late Silurian (Ludlow and Pridoli series).

The base of the Silurian is defined at the base of the *Parakidograptus acuminatus* Zone in the graptolite-rich Dobb's Linn section (GSSP) in Scotland (Cocks 1985). Recent reviews of the development of Silurian rocks in different regions of the world are given by Holland & Bassett (1989) and Bassett et al. (1991). All of the stage boundaries have been formally defined and ratified by the ICS.

With one exception, noted below, no significant new Silurian radiometric data have been published since the thorough timescale review of McKerrow et al. (1985). Therefore, that calibration is adopted here with minor amendments. The age of the base of the Silurian has not been measured, and is estimated at 434 Ma. This estimate is in accord with a SHRIMP zircon date of 430±3 Ma for the Rhuddanian (*Coronograptus cyphus* Zone) Birkhill Shale in Scotland (Compston & Williams 1992). Within the Silurian, the most tightly constrained radiometric age is that of 421±2 Ma for the early Gorstian Laidlaw Volcanics, near Canberra, Australia (Wyborn et al. 1982). This is consistent with an age for the basal Ludlow of close to 420 Ma.

The Silurian was a period of largely cosmopolitan faunas. In Australia, therefore, it has generally been found that stages and even zones (particularly graptolite zones) first recognised in Europe or North America can be reliably applied to isolated successions (Strusz 1989). The only Australian attempt at defining local stages, in Victoria, foundered because it was poorly based both stratigraphically and palaeontologically (Talent et al. 1975).

DEVONIAN PERIOD (410–354 Ma)

The Devonian system of rocks was originally defined and named by Sedgwick & Murchison (1839) for a sequence of marine strata in Devon, which lay beneath Carboniferous strata. Lonsdale's (1840) analysis — that the coral fauna of the limestones in south Devonshire might be post-Silurian and pre-Carboniferous — probably led to the original notion of a Devonian System and the idea that these marine limestones were facies equivalents of the non-marine Old Red Sandstone of north Devon (Rudwick 1979).

The base of the Devonian is defined in the Silurian–Devonian boundary stratotype section (GSSP) at Klonk, near Prague, Czech Republic, at the first appearance of the graptolite *Monograptus uniformis* (see Martinsson 1977). The Devonian Period is divided into Early, Middle and Late series, and eight stages, based on biostratigraphic subdivision in the marine type sections of Europe. All of the stage boundaries have been formally defined and ratified by the ICS.

As with the Silurian, few new radiometric data have been published for the Devonian since the timescale review of McKerrow et al. (1985) (i.e. Rowen et al. 1990; Claoué-Long et al. 1992). The age of the base of the Devonian has not been measured, and Owen & Wyborn's (1979) suggested age of 410 Ma for this level is adopted herein, following Cowie & Bassett (1989). Between this age and the estimated age of 354 Ma for the top of the Devonian (see below), isotopic and biostratigraphic constraints for the interpolation of stage boundaries are generally poor.

In eastern Australia, most of the stage boundaries for the Early and Middle Devonian can be identified in extensive marine sequences (Mawson et al. 1989). The Late Devonian stages are best documented in the Canning and Bonaparte basins of north-western Australia. There are also very extensive successions of non-marine Devonian rocks, particularly in eastern and central Australia (Young 1989).

CARBONIFEROUS PERIOD (354–298 Ma)

The name 'Carboniferous', first used by Conybeare & Phillips (1822) for the 'Coal-measures, Carboniferous limestone, and Old red sandstone' of England and Wales, has the distinction of being the first geological system to be established. The Old Red Sandstone was later removed when the Devonian System was erected (see above).

The base of the Carboniferous is defined by the first appearance of the conodont *Siphonodella sulcata* within the evolutionary lineage from *S. praesulcata* to *S. sulcata*, a level slightly below the lowermost record of the goniatite *Gattendorfia subinvoluta*, previously used as an index macrofossil for this boundary (Paproth 1980). The global stratotype (GSSP) is at La Serre in the Montagne Noire of southern France (Paproth et al. 1991). The Carboniferous is divided into the Mississippian and Pennsylvanian subperiods in North America, and the Dinantian and Silesian subperiods in western Europe.

The age of the base of the Carboniferous has been measured at 353.7$\pm$4.2 Ma for bentonite Bed 79 in the Hasselbachtal auxiliary stratotype section in Germany, 35 cm above the boundary and within the lower part of the *Siphonodella sulcata* conodont zone (Claoué-Long et al. 1992; renormalised by Claoué-Long et al., in press). In eastern Australia, the age for volcanics constrained by the same conodonts is within measurement uncertainty of the Hasselbachtal date. The rest of the Carboniferous timescale is also largely constrained by a new database of zircon U/Pb datings (Claoué-Long et al. 1993, in press; Roberts et al. 1993*a*, *b*, 1995, in press; Riley et al. 1995), and $^{40}Ar/^{39}Ar$ ages (Hess et al. 1983; Lippolt & Hess 1985; Hess & Lippolt 1986), which distinguishes our scale significantly from previous compilations. Our current age estimates for the bases of the western European series boundaries are: 354 Ma Tournaisian; 343 Ma Viséan; 325 Ma Namurian; 312.5 Ma Westphalian; 306 Ma Stephanian.

In general, the Early Carboniferous biota was relatively cosmopolitan. In Australia, therefore, some biochronologic tie-points exist with the Dinantian scale of western Europe, which broadly delimit the standard stages (Jones 1991; Roberts et al. 1993*c*). However, the Late Carboniferous biota of Australia is mainly endemic, as it belongs to the Gondwanan province, outside the palaeoequatorial belt. The rarity of conodonts and the absence of fusulinid foraminiferids inhibit correlation with the standard sequences of North America and Russia, where the main biozonations are based on these groups. This biotal change was associated with major climatic and eustatic changes about the mid-Carboniferous (Mississippian–Pennsylvanian) boundary, resulting from the Late Carboniferous glaciation/ice age. Of the standard Silesian stratigraphic units of western Europe, only the Namurian Series can be recognised with any confidence, on biostratigraphic criteria, in Australian Late Carboniferous sequences. SHRIMP zircon dates from eastern Australia indicate that these sequences may extend at least into the early Westphalian (Roberts et al. 1995).

PERMIAN PERIOD (298–251 Ma)

Murchison (1841) and Murchison et al. (1845) named the Permian System for an extensive sequence, which rests on strata that Murchison thought were of Carboniferous age, in the vicinity of the city of Perm, on the western flank of the Ural Mountains, Russia. However, the Permian System as originally defined by Murchison included only the Kungurian to Tatarian stages; the underlying strata (Artinskian and Sakmarian *sensu lato*) were for a time regarded as Permo–Carboniferous (see Dunbar 1940).

The lower boundary of the Permian is taken to coincide with the base of the fusulinid *Pseudoschwagerina* Zone, and corresponds to the base of the Asselian Stage, now the agreed Russian boundary stratotype for the base of the Permian System in the Urals. It corresponds on the fusulinid scale to the base of the *Sphaeroschwagerina vulgaris–S. fusiformis* Zone (Davydov et al. 1992). Although this position awaits formal international ratification it is widely accepted (Archbold 1982; Kotlyar & Stepanov 1984; Davydov 1988). In western European terms this boundary may be considered equivalent to the Stephanian–Autunian boundary (*sensu* Bouroz & Doubinger 1977). In Australia, the base of the Permian is taken at the appearance of the *Eurydesma*-fauna, which in Western Australia corresponds to the base of the Lyons Group, and in eastern Australia to the base of the Lochinvar Formation (Dickins & Sutherland 1979). The bases of both of these units are markedly transgressive and unconformable. From associated faunal evidence, the plant microfossil *Granulatisporites confluens* Oppel-zone of Foster & Waterhouse (1988) occurs near or at the base of the Permian succession.

The Permian timescale has reasonably closely measured ages for the top and base, but within these limits it lacks reliable ages in areas with satisfactory stratigraphic control. A program of zircon dating is now in progress, pending the outcome of which the ages of the major subdivisions of the Permian are largely conjectural. The age of the base of the Permian is estimated herein at 298 Ma. It is constrained by the 300.3 Ma $^{40}Ar/^{39}Ar$ age reported by Hess et al. (1983) and Lippolt & Hess (1985) for a Stephanian B or C tuff (159/71S) from Baden Baden, Germany, and the 297.8 Ma and 298.7 Ma $^{40}Ar/^{39}Ar$ ages reported by Lippolt & Hess (1983) for tuffs in the Grenzlager Formation (basal Upper Rotliegende) from the Saar–Nahe Basin, Germany. Precise correlation from these non-marine beds to the marine succession of the Russian boundary stratotype is still uncertain.

Except for their use in Tasmania (Clarke & Farmer 1976; Clarke 1989), local stages are not used for the Permian sequences of Australia, which can be correlated with the Russian Ural and Tethyan standards. Provincialism is a distinct feature of Permian marine faunas, but correlation is possible directly and indirectly with the Russian schemes, in many cases with considerable accuracy, especially in parts of Western Australia (Archbold & Dickins 1991). The Taxa used mostly are ammonoids (which are rather sporadic), brachiopods, bivalves, and foraminiferids. Fusulinids are so far entirely absent, and conodonts, present only in Western Australia, are rare. Spore–pollen assemblages span both marine and non-marine sequences; ages throughout the Permian have been determined by comparison of palynofloras (spore–pollen) with dated faunal assemblages.

Mesozoic Era (251–65 Ma)

TRIASSIC PERIOD (251–205 Ma)

The term 'Trias' (subsequently Triassic) was introduced by Alberti (1834) and refers to the threefold lithostratigraphic subdivision of strata (Bunter, Muschelkalk, and Keuper) widely exposed in the Thuringian region of Germany, lithological details of which have been well known since the Middle Ages because of their commercial significance as a source of building stones, halides and gypsum. The sparsely fossiliferous sequences in the original German type area proved difficult to correlate, and the traditional stages of the Triassic were defined in the marine rocks of the Northern Calcareous Alps in Austria.

The boundary between the Permian and the Triassic is marked by a significant extinction event for the world's fauna (Erwin 1994). Curiously, no formal agreement has been reached internationally on the exact placement of this boundary, which also corresponds to the Palaeozoic–Mesozoic boundary, but it is widely accepted that it is at the base of the *Otoceras woodwardi* ammonite zone (Kummel 1972; Tozer 1994; Zakharov 1994). This is the datum used herein.

The base of the Triassic has been dated at the boundary bentonite in the proposed stratotype near Meishan, Changxing, Zhejiang Province, China as 251.1±3.6 Ma by SHRIMP zircon U/Pb dating (Claoué-Long et al. 1991, renormalised by Claoué-Long et al., in press). The bentonite is at the lithological change in the standard section from uppermost Permian Changxing Formation limestone to shales containing mixed faunas that include probable Triassic *Otoceras* ammonoids and, a few centimetres higher, definite Triassic *Hindeodus parvus* conodonts. Tozer (1986) suggested that the bentonite occupies the time interval of a local unconformity, and thus represents a position within the uppermost Permian, rather than its top. Despite this biostratigraphic ambiguity, this age is presently the most direct constraint on the age of the boundary; it is adopted here pending the outcome of new dating studies in the uppermost Permian. Radiometric age control within the Triassic is poor, with many of the control points not adequately constrained biostratigraphically.

In Australia, biochronologic tie-points with the European Triassic standard stages exist in geographically scattered basins and, irregularly and imperfectly, delimit the standard stages (Balme 1991). Ammonites, bivalves, conodonts, nannofossils, dinocysts (cysts of dinoflagellates), and foraminiferids are known from various marine sequences, mainly in the western margin of the continent; and spanning both marine and non-marine sequences are spore–pollen assemblages. Correlations throughout the Triassic have been determined by comparison of palynofloras (spore/pollen/dinocyst) with independently dated faunal assemblages (ammonites and conodonts) from Australia and/or northern hemisphere localities.

JURASSIC PERIOD (205–141 Ma)

The term Jurassic ('Jura Kalkstein') was introduced by Alexander von Humboldt in 1795 for a series of strata exposed in the Jura mountains on the northern border of Switzerland. Studies of the Jurassic in England, France, Germany, and Switzerland, where fossils are abundant, have formulated the scientific principles that govern the present disciplines of lithostratigraphy and biostratigraphy. In 1837 Leopold von Buch introduced the threefold division of the Jurassic that forms the basis for its currently accepted subdivision into three series and eleven stages, which have been defined on ammonites in France, England, and Germany (see Cope et al. 1980*a*, *b*).

The base of the Jurassic, as currently accepted, is at the base of the *Psiloceras planorbis* ammonite zone, the lowermost of three zones that constitute the Hettangian Stage near Hettange in north-eastern France (Cope et al. 1980*a*). In Australia, in the absence of ammonites, this level is taken on palynological evidence to be near the base of the *Corollina torosa* spore–pollen zone, and near the base of the *Dapcodinium priscum* dinocyst zone (Helby et al. 1987).

Very little isotopic control has so far been obtained from the ocean floor. More useful ages (K/Ar, Rb/Sr, Ar/Ar) have been measured in onshore sedimentary and crystalline rocks linked with the fossil record. Armstrong (1982) estimated the Triassic–Jurassic (T–J) boundary as 208 Ma, based on a study of many isotopic analyses of Upper Triassic and Lower Jurassic volcanic rocks in British Columbia, Canada. However, Odin & Letolle (1982) and

Kennedy & Odin (1982) commented on the poor stratigraphic control of several of the Canadian analyses. Quoting additional results from late Triassic intrusives in Thailand (209–211 Ma), the Early Jurassic of the USA (190–206 Ma), and the Hettangian of France (194 Ma), Odin & Odin (1990) deduced an age of 205 Ma for the Triassic–Jurassic boundary, a value accepted in this summary. In the absence of firmer evidence, subdivisions of the Jurassic are here based on equal duration of ammonite zones (on average 1 Ma, assuming a constant rate of evolution).

Although the Jurassic in Australia cannot be correlated with the standard Jurassic stages in Europe directly, attempts to correlate Australian zonations against those stages indirectly, mostly by extrapolation via biogeographic provinces in Laurasia and Gondwana, have been reasonably successful. Non-marine Jurassic rocks are preserved chiefly in north-eastern and central Australia (Great Australian Basin), where their ages are determined by fossil spores and pollen grains (Helby et al. 1987; Filatoff & Price 1988; Burger 1990*a*). Open-sea and restricted marine Jurassic rocks are known chiefly from Western Australia (coastal onshore basins and North West Shelf), where they are zoned on the evidence from spores, pollen grains, and marine microfossils (dinocysts, foraminiferids), and calcareous nannofossils (Shafik 1994). Ammonites are present at certain intervals, where they contribute to biozonal determination (Helby et al. 1987; Hall 1989).

CRETACEOUS PERIOD (141–65 Ma)

The term Cretaceous (*Terrain Crétacé*) was introduced in 1822 by J.B. d'Omalius d'Halloy for strata encircling the Paris Basin. The name is from the Latin *creta*, meaning chalk, the most conspicuous rock type in the system in north-western Europe. Alcide D'Orbigny first grouped Cretaceous fossil assemblages into what he termed '*étages*', thus forming the basis of the present subdivision into two series (Lower and Upper Cretaceous) and twelve stages, defined mainly on ammonites in France, Switzerland, and the Netherlands (Birkelund et al. 1984; Kennedy 1984).

The base of the Cretaceous is traditionally accepted as the base of the *Berriasella grandis* ammonite zone, the lowermost of three zones that constitute the Berriasian Stage near Berrias in eastern France. Proposals are being investigated to establish the base of the Cretaceous at the base of the preceding ammonite zone (*Berriasella jacobi* Zone), whose equivalence is recognised outside the Tethyan region in Europe (Colloque sur la limite Jurassique–Crétacé 1975). In Australia, palynology provides the most detailed fossil record of Jurassic–Cretaceous (J–K) boundary strata; the base of the Cretaceous is accepted as lying within the lower *Ruffordiaspora australiensis* and *Crybelosporites stylosus* spore–pollen zones, and within the *Kalyptea wisemaniae* dinocyst zone (Dettmann et al. 1992; Burger, in press).

There are few reliable age indicators for the base of the Cretaceous at the base of the *B. grandis* Zone. Kennedy & Odin (1982) cited glauconite datings of 122 and 134 Ma for the Ryazanian (Berriasian–Valanginian) in England and Russia. Harland et al. (1982) calculated 135 Ma for the Berriasian–Valanginian boundary on the basis of minimum error functions from dated glauconite, but by the same method Lowrie & Ogg (1986) obtained 135 Ma for the J–K boundary (base of *B. grandis* Zone). However, the estimate of 141 Ma by Bralower et al. (1990) is based on the closest reliable date to the J–K boundary, a zircon U/Pb age of 137.1±0.6 Ma for a horizon within the Berriasian. The J–K boundary age of 141 Ma is based on close extrapolation of the precise zircon U/Pb age with the nannofossil and *Buchia* (Bivalvia) biozonations, and the geomagnetic polarity timescale. At present no accurate ages have been obtained for the Early Cretaceous sea-floor anomalies. The timescale of Gradstein et al. (1994) was not formally available for comparison at the time of this compilation. Thus, our timescale follows those of Haq & Van Eysinga (1987) for the Early Cretaceous, slightly modified to accept 141 Ma for the J–K boundary (see above), and Harland et al. (1982) for the Late Cretaceous. Few acceptable analyses are available for Cretaceous Stage boundaries, these being mainly confined to the Late Cretaceous. On the basis of radiometrically dated Palaeogene and Late Cretaceous ocean-floor anomalies, Tarling & Mitchell (1976), Lowrie & Alvarez (1981), Harland et al. (1982) and Berggren et al. (1985) have extrapolated closely comparable age estimates for Late Cretaceous anomalies 29 to 34 (Maastrichtian, Campanian).

The Cretaceous in Australia, like the Jurassic, cannot be correlated directly with the type sequences in Europe. However, the ammonite record, although incomplete, is indicative of certain stages, and attempts to correlate regional faunal and floral zonal schemes against the standard Cretaceous, by correlation via biogeographic provinces in Laurasia and Gondwana, have been largely successful (Burger 1990*b*). Cretaceous strata have been widely mapped and studied on the Australian continent, and along the southern, western, and northern margins. Both non-marine and shallow marine strata occur in eastern Australia (Great Australian and Otway–Gippsland Basins, Bass Strait); they are zoned by spores, pollen, megaplants, and marine fossils (dinocysts, ammonites, foraminiferids and calcareous nannofossils; Day 1969; Burger 1973, 1982; Haig 1979; Morgan 1980; Shafik 1985; Dettmann et al. 1992). The most complete marine sequences (shelf-facies) known are in Western Australia (Perth, Carnarvon, Canning Basins) and on the North West Shelf, where they are zoned chiefly by dinocysts, spores, pollen, foraminiferids, ammonites, and calcareous nannofossils (Helby et al. 1987; McNamara et al. 1988; Henderson 1990; Shafik, 1990; Henderson et al. 1992; Burger, in press).

Cainozoic Era (65–0 Ma)

TERTIARY PERIOD (65–1.78 Ma)

The term Tertiary is the only one of three main categories of rocks (Primary, Secondary, Tertiary) introduced by Giovanni Arduino in 1759 that is still in use. It was originally defined in Italy and was applied by Brongniart in 1810 to strata overlying the Cretaceous chalk in the Paris Basin. All of the series/epochs of the Tertiary (Paleocene, Eocene, Oligocene, Miocene and Pliocene) have their stratotypes in France. Charles Lyell (1833) defined the Eocene, Miocene and Pliocene series/epochs based on studies of the fossil molluscs by himself and Deshayes, and in particular on the percentage of species in common with the modern fauna. August von Beyrich (1854) later added the Oligocene Series, and Schimper (1874) introduced the Paleocene Series. Present stratigraphic usage divides the Tertiary into the Palaeogene and Neogene subperiods (or periods, if the Tertiary is regarded as a sub-era, after Harland et al. 1990). The Palaeogene includes three epochs, Paleocene, Eocene, Oligocene; the Neogene two epochs, Miocene and Pliocene.

The base of the Palaeogene (the base of the Danian Stage), and hence the base of the Tertiary, has been defined by international agreement in the El Haria section, near El Kef, Tunisia (Jenkins & Luterbacher 1992; Odin 1992). It can be delimited by the sharp turnover in the marine planktic assemblages: the mass extinction of most of the Cretaceous biota and the emergence of new forms. The Cretaceous–Tertiary (K–T) boundary is marked by the extinction of the Cretaceous planktic foraminiferids *Globotruncana* and *Rugoglobigerina* (Blow 1979; Berggren et al. 1985, in press) and the coccoliths *Micula* spp., *Nephrolithus* spp., *Arkhangelskiella* spp., and others, and approximately (but not precisely) by the first appearance of the foraminiferid *Parvularugoglobigerina eugubina* (Berggren et al., in press) and the acme of *Thoracosphaera* (Perch-Nielsen 1985). The base of the Tertiary has not yet been recognised in Australia, due to a hiatus spanning the latest Maastrichtian and early Danian. Zone P1c foraminiferid faunas are the oldest known.

The age of the K–T boundary is now taken at 65 Ma, principally from $^{40}Ar/^{39}Ar$ dating of tektites in the boundary event layer, thought to be associated with the impact represented by the Chicxulub crater (Swisher et al. 1992; Dalrymple et al. 1993; Berggren et al., in press). Berggren et al. (1993) discuss how variations on this age in the literature are mainly a function of intercalibration problems in the $^{40}Ar/^{39}Ar$ dating method, and show that the weighted mean of identically normalised $^{40}Ar/^{39}Ar$ ages is 65.06 ± 0.02 Ma. The latest available age information for subdivisions of the Tertiary is discussed by Berggren et al. (in press), whose reviews are followed in constructing this scale, and to which the reader is referred for detailed discussions of the age calibration. This section of the timescale is becoming very accurately constrained from a combination of precise $^{40}Ar/^{39}Ar$ dating and biostratigraphic interpolations securely tied to the magnetostratigraphic record. In the youngest parts of the scale, astronomical calibrations are helping to provide an independent check on radiometric age measurements. The Paleocene–Eocene boundary is constrained at 54.8 Ma by $^{40}Ar/^{39}Ar$ dating of ashes at this level in Denmark, and correlatives in England. The calibration of the Eocene–Oligocene boundary at 33.7 Ma (Odin 1989) is thoroughly discussed by Berggren et al. (1992), Odin (1992) and Odin & Luterbacher (1992), and supported by new data discussed in Berggren et al. (in press). The Oligocene–Miocene boundary is taken at 23.8 Ma, an age agreed on by both Berggren et al. (in press) and Harland et al. (1990). The top of the Miocene is placed at 5.32 Ma, significantly older than the age used in previous compilations by Cande & Kent (1992: 5.16Ma) and Berggren et al. (1985: 4.86 Ma), partly on the basis of astronomical calibration (Hilgen 1991*a*). Astronomical calibration has also helped to clarify the $^{40}Ar/^{39}Ar$ age estimates for the Pliocene–Pleistocene boundary at 1.78 Ma (Hilgen 1991*a*).

Tertiary marine sediments are mainly confined to the margins and offshore areas of Australia, whereas non-marine sediments have a broader, continental distribution. The onshore marine deposits tend to be of shallow-water origin, which compounds problems resulting from the often most important palaeoceanic and palaeolatitudinal factors affecting the distribution of marine calcareous planktic fossils. Correlation with global zonal schemes has often been difficult, especially for schemes based on extratropical planktic and nannofossil events in the southern Australian sequence (Shafik 1983; McGowran 1989). Within the warmer sections, global zonal schemes based on marine calcareous plankton have been applied successfully, although some local modification has been necessary. Local marine calcareous plankton zonal schemes are used mainly in the temperate sequences of southern Australia, mainly because there are difficulties in applying the global schemes to this area. A local stage scheme (Carter 1958) is also available for this area, but it has had little acceptance, especially in recent years, because it is based mainly on benthic assemblages, with the consequent difficulties in correlation with global schemes, and also because of temporal uncertainties of the relationship between these stages. For example, there is good evidence that the Batesfordian and Balcombian are time correlatives, at least in part (Carter 1990). European stage nomenclature has also been used, mainly by petroleum exploration companies. Tertiary epochs (Paleocene to Pliocene) are recognised and widely used in association with the use of zones/bioevents of the main groupings, planktic foraminiferids, calcareous nannofossils and palynomorphs. Spore–pollen zonations reflect an Australian vegetation clearly different from that found elsewhere, and are based on reference sections in south-eastern Australia (Truswell et al. 1991). Shallow-water marine carbonates from the northern parts of Australia often contain larger benthic foraminiferal assemblages, enabling the use of the East Indian Letter Stage system, particularly for the late Eocene, and late Oligocene to middle Miocene.

QUATERNARY PERIOD (1.78–0 Ma)

In 1829 Desnoyers classified some marine, lacustrine and alluvial sediments from the Seine Basin under the name 'Quaternary' because he considered them to be younger than the Tertiary rocks. Reboul, in 1833, discussed the Quaternary Period and noted that it was characterised by fossils very similar to living species. He considered that the Quaternary had two subdivisions: a historical part and a prehistoric, post-Tertiary part. At about the same time it was realised that the widespread erratics in the Alps and northern Europe were the result of deposition from extensive ice sheets. In 1839 Lyell proposed the Pleistocene Period for this 'ice age', which post-dated his established Pliocene of the Tertiary and preceded his Recent. In 1854 Morlot followed Reboul's usage of Quaternary, but recognised the glacial sequences as part of the division.

The base of the Pleistocene, and hence the base of the Quaternary, was defined by the ICS in the Vrica section near Crotone, Calabria, in Italy. The boundary, that is, the base of the Calabrian Stage, lies above the last appearance datum (LAD) of the nannofossil *Discoaster broweri*, and below the first appearance datum (FAD) of the nannofossil *Geophyrocapsa oceanica* and the foraminiferid *Globigerinoides tenellus*. However, the Pliocene–Pleistocene boundary as defined at the Vrica section is difficult to recognise on planktic foraminiferal faunas in deep-sea localities and areas outside the type area. The first evolutionary appearance of the planktic foraminiferid *Globorotalia truncatulinoides* was considered a satisfactory criterion to recognise this boundary (Hays & Berggren 1971), but this datum is now known to be diachronous (Dowsett 1989). To overcome this problem, Berggren et al. (in press) have suggested that the last occurrence of *Globigerinoides fistulosus* be used to mark this boundary (between their Zone P16 and Zone N22) in the SW Pacific and Indian Oceans, and the last occurrence of *Globigerinoides extremus* in low latitude areas.

At Vrica, the Pliocene–Pleistocene boundary is placed within the upper part of the Olduvai Subchron, according to magnetostratigraphic studies by Hilgen (1991*b*) and Zijderveld et al. (1991). Hilgen (1991*b*) reported an age of 1.81 Ma for this boundary in the stratotype section, based on astronomical calibrations using sapropels. Also, new palaeomagnetic studies at this section suggest that the position of the top of the Olduvai magnetic event is ambiguous. Further, Hilgen (1991*b*) does not relate the key biostratigraphic events (LADs of *Globigerinoides fistulosus* and *G. extremus*) to his new age. Because of the uncertainities of correlation between the position of the Pliocene–Pleistocene boundary with both magnetostratigraphic and biostratigraphic events, we prefer, at this stage, to use the previous age estimate for this boundary of 1.78 Ma.

As for the Tertiary, onshore marine sequences for the Quaternary are of shallow-water origin, and so contain few planktic species, whereas offshore sections often contain rich and diverse planktic assemblages. These offshore sequences could be studied only recently, due to modern dredging and coring operations. To date few studies have been made, but global zones can readily be recognised, and new subdivisions of these have been established for north-eastern Australia (Chaproniere 1991). The onshore areas are covered in a veneer of non-marine sediments; pollen analyses are concentrated in the late Pleistocene and have, as their focus, vegetation and climatic reconstruction. For south-eastern Australia two local stage names have been applied to the Quaternary, but these have found little acceptance due to the difficulties in understanding temporal and spatial relationships between them (Carter 1990). Traditional epoch nomenclature (Pleistocene and Holocene) has been applied to the Australian region, but biostratigraphic zonal nomenclature is gaining wide acceptance.

ACKNOWLEDGMENTS

As the compiler of this chart and accompanying notes, I take pleasure in thanking my AGSO colleagues for their assistance in many ways. Their work is cited in the list of references. I have drawn heavily on the data compiled by Gavin C. Young and John R. Laurie in their work, *An Australian Phanerozoic Timescale.* John H. Shergold encouraged me to write this summary in order to distribute the chart as widely as possible. Finally, I am particularly indebted to Lynton Spencer for entering the draft chart into a database that can be manipulated to produce a variety of versions for publication.

MEMBERS OF AGSO TIMESCALE CALIBRATION AND DEVELOPMENT PROJECT

Burger, D.	[Jurassic, Cretaceous]
Chaproniere, G.C.H.	[Cainozoic]
Claoué-Long, J.C.	[Geochronology]
Dickins, J.M.	[Permian]
Foster, C.B.	[Triassic]
Jones, P.J.	[Devonian, Carboniferous]
Laurie, J.R.	[Ordovician]
Nicoll, R.S.	[Cambrian to Triassic]
Shafik, S.	[Jurassic, Cretaceous, Cainozoic]

Shergold, J.H.*	[Cambrian, Ordovician]
Strusz, D.L.	[Silurian]
Truswell, E.M.	[Cainozoic]
Young, G.C.	[Devonian]

* Project leader

AUSTRALIAN ADVISORS TO THE AGSO TIMESCALE CALIBRATION AND DEVELOPMENT PROJECT

Apthorpe, M.C.	[Cretaceous, Cainozoic] — Apthorpe Palaeontology Pty Ltd, Perth
Archbold, N.W.	[Permian] — Deakin University, Victoria
Balme, B.E.	[Triassic] — University of Western Australia
Cooper, B.J.	[Cambrian] — Mines & Energy South Australia
Gravestock, D.I.	[Cambrian] — Mines & Energy South Australia
Jago, J.B.	[Cambrian] — University of South Australia
Jenkins, T.B.H.	[Carboniferous] — Sydney University
McGowran, B.	[Cainozoic] — University of Adelaidae
Macphail, M.K.	[Cainozoic] — Australian National University
Mawson, R.	[Devonian] — Macquarie University, Sydney
Partridge, A.D.	[Cainozoic] — La Trobe University, Victoria
Playford, G.	[Ordovician] — University of Queensland
Plumb, K.A.	[Precambrian] — AGSO, Canberra
Roberts, J.	[Carboniferous] — University of NSW, Sydney
Stait, B.	[Ordovician] — Parliamentary Library, Hobart, Tasmania
Talent, J.A.	[Devonian] — Macquarie University, Sydney
VandenBerg, A.H.M.	[Ordovician] — Department of Agriculture, Energy & Minerals, Victoria
Webby, B.D.	[Ordovician] — Macquarie University, Sydney

References

Alberti, F.A., von, 1834. *Beitrag zu einer Monographie des Bunten Sandsteins, Muschelkalks und Keuper, und die Verbindung dieser Gebilde zu einer Formation*, J.G. Cotta, Stuttgart.

Archbold, N.W., 1982. Correlation of the Early Permian faunas of Gondwana: Implications for the Carboniferous–Permian boundary. *Journal of the Geological Society of Australia* 29(3 & 4), 267–75.

Archbold, N.W. & Dickins, J.M., 1991. Australian Phanerozoic Timescales. 6. Permian. A standard for the Permian System in Australia. *Bureau of Mineral Resources, Geology & Geophysics, Record* 1989/36, 17 pp.

Armstrong, R.L., 1982. Late Triassic–Early Jurassic time-scale calibration in British Columbia, Canada. pp. 509–13 *in* G.S. Odin (ed.), *Numerical Dating in Stratigraphy Part I.* John Wiley & Sons Ltd, Chichester.

Balme, B.E., 1991. Australian Phanerozoic Timescales. 7. Triassic. Biostratigraphic Chart and Explanatory Notes, *Bureau of Mineral Resources, Geology & Geophysics, Record* 1989/37, 28 pp.

Bassett, M.G., Lane, P.D. & Edwards, D. (eds), 1991. *The Murchison Symposium: proceedings of an international conference on The Silurian System.* Special Papers in Palaeontology 44, 1–397.

Becker, R.T., House, M.R., & Kirchgasser, W.T., 1993. Devonian goniatite biostratigraphy and timing of facies movements in the Frasnian of the Canning Basin, Western Australia. pp. 293–321 *in* E.A. Hailwood & R.B. Kidd (eds), *High Resolution Stratigraphy*, Geological Society Special Publication 70.

Berggren, W.A., Kent, D.V., & Flynn, J.J., 1985. Paleogene geochronology and chronostratigraphy. pp. 141–95 *in* N.J. Snelling (ed.), *The Chronology of the Geological Record*, Blackwell, London.

Berggren, W.A., Kent, D.V., Obradovitch, J.D. & Swisher, C.C., III, 1992. Towards a revised Paleogene geochronology. pp. 29–45 *in* D.R. Prothero & W.A. Berggren (eds), *Eocene–Oligocene Climatic and Biotic Evolution*, Princeton University Press, Princeton, NJ.

Berggren, W.A., Kent, D.V., Swisher C.C. III & Aubry, M.-P., in press. A revised Cenozoic geochronology and chronostratigraphy. *In* W.A. Berggren, D.V. Kent, & J. Hardenbol (eds), *Geochronology, Time Scales and Global Stratigraphic Correlations: a Unified Temporal Framework for an Historical Geology*, Society of Economic Paleontologists and Mineralogists Special Publication.

Beyrich, von E., 1854. Über die Stellung der Hessischen Tertiärbildungen. *Berichte über die Verhandlungen der Königlichen Preussischen Akademie der Wissenschaften, Berlin 1854*, 640–66.

Birkelund, T., Hancock, J.M., Hart, M.B., Rawson, P.F., Remane, J., Robaszynski, F., Schmid, F. & Surlyk, F., 1984. Cretaceous stage boundaries — proposals. *Geological Society of Denmark Bulletin* 33, 3–20.

Blow, W.H., 1979. *The Cainozoic Globigerinida. A study of the Morphology, Taxonomy, Evolutionary Relationships and the Stratigraphical Distribution of Some Globigerinida (mainly Globigerinacea)*, E.J. Brill, Leiden.

Bouroz, A. & Doubinger, J., 1977. Report on the Stephanian–Autunian boundary and on the contents of Upper Stephanian and Autunian in their stratotypes. pp. 147–69 *in* V.M. Holub & R.H. Wagner (eds), Symposium on Carboniferous Stratigraphy, *Ústředni ústav Geologický.*

Bowring, S.A., Grotzinger, J.P., Isachsen, C.E., Knoll, A.H., Pelechaty, S.M. & Kolosov, P., 1993. Calibrating rates of early Cambrian evolution. *Science* 261, 1293–8.

Bralower, T.J., Ludwig, K.R., Obradovich, J.D. & Jones, D.L., 1990. Berriasian (Early Cretaceous) radiometric ages from the Grindstone Creek section, Sacramento Valley, California. *Earth and Planetary Science Letters* 98, 62–73.

Burger, D., 1973. *Spore zonation and sedimentary history of the Neocomian, Great Artesian Basin, Queensland.* Geological Society of Australia Special Publication 4, 87–118.

Burger, D., 1982. A basal Cretaceous dinoflagellate suite from northeastern Australia. *Palynology* 6, 161–92.

Burger, D., 1990*a*. Australian Phanerozoic Timescales. 8. Jurassic Biostratigraphic Chart and Explanatory Notes. *Bureau of Mineral Resources, Geology & Geophysics, Record* 1989/38, 32 pp.

Burger, D., 1990*b*. Australian Phanerozoic Timescales. 9. Cretaceous Biostratigraphic Chart and Explanatory Notes. *Bureau of Mineral Resources, Geology & Geophysics, Record* 1989/39, 35 pp.

Burger, D., in press. Cretaceous. *In* G.C. Young & J.R. Laurie (eds), *An Australian Phanerozoic Timescale*, Oxford University Press, Melbourne.

Campbell, K.S.W., 1990. Palaeontological contributions to modern evolutionary theory: 1986 Mawson Lecture. *Australian Journal of Earth Sciences* 37(3), 247–65.

Cande, S. & Kent, D.V., (1992). A new geomagnetic polarity time scale for the Late Cretaceous and Cenozoic. *Journal of Geophysical Research* 97, 13 917–51.

Carter, A.N., 1958. Pelagic foraminifera in the Tertiary of Victoria. *Geological Magazine* 95, 297–304.

Carter, A.N., 1990. Time and space events in the Neogene of south-east Australia. pp. 183–93 *in* R. Tsuchi (ed.), *Pacific Neogene Events, Their Timing, Nature and Interrelationship*, University of Tokyo Press, Toyko.

Chadwick, G.H., 1930. Subdivision of geologic time (abstract). *Geological Society of America Bulletin* 41, 47.

Chaproniere, G.C.H., 1991. Pleistocene to Holocene planktic foraminiferal biostratigraphy of the Coral Sea, offshore Queensland, Australia. *BMR Journal of Australian Geology & Geophysics* 12(3), 195–221.

Claoué-Long, J.C., Compston, W., Roberts, J. & Fanning, M., in press. Two Carboniferous ages: a comparison of SHRIMP zircon dating with conventional zircon ages and $^{40}Ar/^{39}Ar$ analysis. *In* W.A. Berggren, D.V. Kent & J. Hardenbol (eds), *Geochronology, Time Scales and Global Stratigraphic Correlation: a Unified Temporal Framework for an Historical Geology*, Society of Economic Paleontologists and Mineralogists Special Publication.

Claoué-Long, J.C., Jones, P.J. & Roberts, J., 1993. The age of the Devonian–Carboniferous boundary. pp. 531–49 *in* M. Streel, G. Sevastopulo & E. Paproth (eds), Devonian–Carboniferous boundary. *Annals de la Société Géologique de Belgique*, 115(2).

Claoué-Long, J.C., Jones, P.J., Roberts, J. & Maxwell, S., 1992. The numerical age of the Devonian–Carboniferous boundary. *Geological Magazine* 129(3), 281–91.

Claoué-Long, J.C., Zhang Zichao, Ma Guogan & Du Shaohua, 1991. The age of the Permian–Triassic boundary. *Earth and Planetary Science Letters* 105, 182–90.

Clarke, M.J., 1989. Lower Parmeener Supergroup. pp. 295–309 *in* C.F. Burrett & E.L. Martin (eds), *Geology and mineral resources of Tasmania*, Geological Society of Australia Special Publication 15.

Clarke, M.J. & Farmer, N., 1976. Biostratigraphic nomenclature for Late Palaeozoic rocks in Tasmania. *Papers and Proceedings of the Royal Society of Tasmania* 110, 91–109.

Cloud, P. & Glaessner, M.F., 1982. The Ediacarian Period and System: Metazoa inherit the Earth. *Science* 217(4562), 783–92.
Cocks, L.R.M., 1985. The Ordovician–Silurian boundary. *Episodes* 8(2), 98–100.
Colloque sur la limite Jurassique–Crétacé (Lyon & Neuchâtel 1973), 1975. Mémoire du Bureau de Recherches Géologiques et Minières 86, 393 pp.
Compston, W. & Pidgeon, R.T., 1986. Jack Hills, evidence of more very old detrital zircons in Western Australia. *Nature* 321, 766–9.
Compston, W. & Williams, I.S., 1992. Ion probe ages for the British Ordovician and Silurian stratotypes. pp. 59–67 *in* B.D. Webby & J.R. Laurie (eds), *Global Perspectives on Ordovician Geology*, Balkema, Rotterdam.
Conway Morris, S., 1989. South-eastern Newfoundland and adjacent areas (Avalon Zone). pp. 7–39 *in* J.W. Cowie & M.D. Brasier (eds), *The Precambrian–Cambrian Boundary*, Clarendon Press, Oxford.
Conybeare, W.D. & Phillips, W., 1822. *Outlines of the Geology of England and Wales, with an Introductory Compendium of the General Principles of that Science, and Comparative Views of the Structure of Foreign Countries*, Part I. William Phillips, London, lxi+470 pp.
Cooper, R.A. & Lindholm, K., 1990. A precise worldwide correlation of early Ordovician graptolite sequences. *Geological Magazine* 127(6), 497–525.
Cope, J.C.W., Getty, T.A., Howarth, M.K., Morton, N. & Torrens, H.S., 1980*a*. *A correlation of Jurassic rocks in the British Isles. Part One: Introduction and Lower Jurassic.* Geological Society of London Special Report 14, 73 pp.
Cope, J.C.W., Duff, K.L., Parsons, C.F., Torrens, H.S., Wimbledon, W.A. & Wright, J.K., 1980*b*. *A correlation of Jurassic rocks in the British Isles. Part Two: Middle and Upper Jurassic.* Geological Society of London Special Report 15, 109 pp.
Cowie, J.W. & Bassett, M.G., 1989. IUGS 1989 Global Stratigraphic Chart with geochronometric and magnetostratigraphic calibration. Supplement to *Episodes* 12(2).
Dalrymple, G.B., Izett, G.A., Snee, L.W. & Obradovich, J.D., 1993. $^{40}Ar/^{39}Ar$ age spectra and total-fusion ages of tektites from Cretaceous–Tertiary Boundary sedimentary rocks in the Beloc Formation, Haiti. *U.S. Geological Survey Bulletin* 2065, 1–20.
Davydov, V.I., 1988. On the problems and principles of stage and zonal division of the Upper Carboniferous. *Annual of the All-Union Paleontological Society, Leningrad* 31, 256–69.
Davydov, V.I., Barskov, I.S., Bogoslovskaya, M.F., Leven, E.Y., Popov, A.V., Akhmetshina, L.Z. & Kozitskaya, R.I., 1992. The Carboniferous–Permian Boundary in the former USSR and its correlation. *International Geology Review* 34(9), 889–906.
Day, R.W., 1969. The Lower Cretaceous of the Great Artesian Basin. pp. 140–73 *in* K.S.W. Campbell (ed.), *Stratigraphy and Palaeontology. Essays in honour of Dorothy Hill*, Australian National University Press, Canberra.
Dettmann, M.E., Molnar, R.E., Douglas, J.G., Burger, D., Fielding, C., Clifford, H.T., Francis, J., Jell, P., Rich, T., Wade, M., Rich, P.V., Pledge, N., Kemp, A. & Rozefelds, A., 1992. Australian Cretaceous terrestrial faunas and floras: biostratigraphic and biogeographic implications. *Cretaceous Research* 13, 207–62.
Dickins, J.M. & Sutherland, W.A., 1979. A glacial moraine in the Permian Lochinvar Formation of the Sydney Basin, New South Wales. *Search* 10(5), 190–1.
Dowsett, H.I., 1989. Application of the graphic correlation method to Pliocene marine sequences. *Marine Micropaleontology* 14, 3–32.
Dunbar, C.O., 1940. The type Permian: its classification and correlation. *American Association of Petroleum Geologists Bulletin* 24, 237–81.

Elles, G.L., 1925. The characteristic assemblages of the graptolite zones of the British Isles. *Geological Magazine* 62, 337–47.
Erwin, D.H., 1994. The Permo–Triassic extinction. *Nature* 367, 231–6.
Filatoff, J. & Price, P.L., 1988. A pteridacean spore lineage in the Australian Mesozoic. *Association of Australasian Palaeontologists Memoir* 5, 89–124.
Fortey, R.A., Harper, D.A.T., Ingham, J.K., Owen, A.W. & Rushton, A.W.A., 1995. A revision of Ordovician series and stages from the historical type area. *Geological Magazine* 132 (1), 15–30.
Foster, C.B. & Waterhouse, J.B., 1988. The *Granulatisporites confluens* Oppel-zone and Early Permian marine faunas from the Grant Formation on the Barbwire Terrace, Canning Basin, Western Australia. *Australian Journal of Earth Science* 35(2), 135–57.
Froude, D.O., Ireland, T.R., Kinny, P.D., Williams, I.S., Compston, W., Williams, I.R. & Myers, J.S., 1983. Ion microprobe identification of 4100–4200 Myr-old terrestrial zircons. *Nature* 304(5927), 616–18.
Glaessner, M.F., 1984. *The Dawn of Animal Life: A Biohistorical Study*, Cambridge University Press, Cambridge, xi + 244 pp.
Glenister, B.F. & Klapper, G., 1966. Upper Devonian conodonts from the Canning Basin, Western Australia. *Journal of Paleontology* 40(4), 777–842.
Gradstein, F.M., Agterberg, F.P., Ogg, J.G., Hardenbol, J., Van Veen, P., Thierry, J. & Huang, Z., 1994. A Mesozoic time scale. *Journal of Geophysical Research* 99(B12), 24 051–74.
Grey, K., 1994. Ediacarian acritarch biostratigraphy of the Centralian Superbasin. *Australian Palaeontological Convention, February 7–9, 1994, Abstracts*, 32.
Haig, D.W., 1979. Cretaceous foraminiferal biostratigraphy of Queensland. *Alcheringa* 3, 171–87.
Hall, R.L., 1989. Lower Bajocian ammonites (Middle Jurassic; Soninniidae) from the Newmarracarra Limestone, Western Australia. *Alcheringa* 13, 1–20.
Hall, T.S., 1897. Victorian Graptolites. Part I. Ordovician from Matlock. (b) *Dictyonema macgillivrayi*, nom. mut. *Proceedings of the Royal Society of Victoria*, 10, 13–16.
Haq, B.U. & van Eysinga, F.W.B., 1987. *Geological Time Table* (4th edn, Wall Chart), Elsevier, Amsterdam.
Harland, W.B., Armstrong, R.L., Cox, A.V., Craig, L.E., Smith, A.G. & Smith, D.G., 1990. *A Geologic Time Scale 1989*, Cambridge University Press, Cambridge, 263 pp.
Harland, W.B., Cox, A.V., Llewellyn, P.G., Pickton, C.A.G., Smith, A.G. & Walters. R., 1982. *A Geologic Time Scale*, Cambridge University Press, Cambridge, 131 pp.
Hays, J.D. & Berggren, W.A., 1971. Quaternary boundaries and correlations. pp. 669–91 *in* B.M. Funnell & W.R. Riedel (eds), *The Micropalaeontology of Oceans*, Cambridge University Press, Cambridge.
Helby, R., Morgan, R. & Partridge, A.D., 1987. A palynological zonation of the Australian Mesozoic. *Association of Australasian Palaeontologists, Memoir* 4, 1–94.
Henderson, R.A., 1990. Late Albian ammonites from the Northern Territory, Australia. *Alcheringa* 14(2), 109–48.
Henderson, R.A., Kennedy, W.J. & McNamara, K.J., 1992. Maastrichtian heteromorph ammonites from the Carnarvon Basin, Western Australia. *Alcheringa* 16(2), 133–70.
Hess, J.C., Backfisch, S. & Lippolt, H.J. 1983. Konkordantes Sanidin- und diskordante Biotitalter eines Karbontuffs der Baden-Badener Senke, Nordschwarzwald. *Neues Jahrbuch für Geologie und Paläontologie Monatshefte* 1983(5), 277–92.
Hess, J.C. & Lippolt, H.J., 1986. $^{40}Ar/^{39}Ar$ ages of tonstein and tuff sanidines: new calibration points for the improvement of the Upper Carboniferous time scale. *Chemical Geology (Isotope Geoscience Section)* 59, 143–54.

Hilgen, F.J., 1991*a*. Extension of the astronomically calibrated (polarity) time scale to the Miocene/Pliocene boundary. *Earth and Planetary Science Letters* 107, 349–66.

Hilgen, F.J., 1991*b*. Astronomical calibration of Gauss to Matuyama sapropels in the Mediterranean and implication for the Geomagnetic Polarity Time Scale. *Earth and Planetary Science Letters* 104, 226–44.

Holland, C.H. & Bassett, M.G. (eds), 1989. *A global standard for the Silurian System*, National Museum of Wales, Geological Series 9, Cardiff, 325 pp.

Huff, W.D. & Kolata, D.R., 1990. Correlation of the Ordovician Deicke and Millbrig K-bentonites between the Mississippi Valley and the southern Appalachians. *American Association of Petroleum Geologists Bulletin* 74, 1736–47.

Isachsen, C.E., Bowring, S.A., Landing, E. & Samson, S.D., 1994. New constraint on the division of Cambrian time. *Geology* 22, 496–8.

Jaanusson, V., 1960. On the Series of the Ordovician System. *21st Session of the International Geological Congress, Norden 1960, Reports VII*, 70–81.

Jenkins, D.G. & Luterbacher, H.-P., 1992. Paleogene stages and their boundaries (Introductory remarks). *Neues Jahrbuch für Geologie und Paläontologie Abhandlungen* 186(1/2), 1–5.

Jenkins, R.F., 1981. The concept of an 'Ediacaran Period' and its stratigraphic significance in Australia. *Transactions of the Royal Society of South Australia* 105(4), 179–94.

Jenkins, R.F., 1990. The Adelaidean Fold Belt: Tectonic reappraisal. pp. 398–420 *in* J.B. Jago & P.S. Moore (eds), *The evolution of a Late Precambrian–Early Palaeozoic Rift Complex: The Adelaide Geosyncline.* Geological Society of Australia Special Publication 16.

Jones, P.J., 1991. Australian Phanerozoic Timescales. 5. Carboniferous Biostratigraphic Chart and Explanatory Notes. *Bureau of Mineral Resources, Geology & Geophysics, Record* 1989/35, 43 pp.

Jones, P.J., Shergold, J.H. & Druce, E.C., 1971. Late Cambrian and Early Ordovician stages in western Queensland. *Journal of the Geological Society of Australia* 18(1), 1–32.

Keller, B.M., 1979. Precambrian stratigraphic scale of the USSR. *Geological Magazine* 116, 419–29.

Kennedy, W.J., 1984. Ammonite faunas and the 'standard zones' of the Cenomanian to Maastrichtian Stages in their type areas, with some proposals for the definition of the stage boundaries by ammonites. *Geological Society of Denmark Bulletin* 33, 147–61.

Kennedy, W.J. & Odin, G.S., 1982. The Jurassic and Cretaceous time scale in 1981. pp. 557-92 *in* G.S. Odin (ed.), *Numerical dating in Stratigraphy, Part 1*, John Wiley & Sons Ltd, Chichester.

Kotlyar, G.V. & Stepanov, D.L. (eds), 1984. Main features of stratigraphy of the Permian System in the USSR. *A.P. Karpinsky All-Union Order of Lenin Geological Research Institute, Transactions, new series* 286, 1–280.

Kummel, B., 1972. The Lower Triassic (Scythian) ammonoid *Otoceras*. *Bulletin of the Museum of Comparative Zoology, Harvard University* 143(6), 365–417.

Landing, E., 1994. Precambrian–Cambrian boundary global stratotype ratified and a new perspective of Cambrian time. *Geology* 22, 179–82.

Lapworth, C., 1879. On the tripartite classification of the Lower Palaeozoic rocks. *Geological Magazine* 16, 1–15.

Lippolt, H.J. & Hess, J.C., 1983. Isotopic evidence for the stratigraphic position of the Saar-Nahe Rotliegend volcanism I. $^{40}Ar/^{40}K$ and $^{40}Ar/^{40}Ar$ investigations. *Neues Jahrbuch für Geologie und Paläontologie Monatshefte* 1983(12), 713–30.

Lippolt, H.J. & Hess, J.C., 1985. Ar^{40}/Ar^{39} dating of sanidines from Upper Carboniferous tonsteins. *Compte rendu 10 -ième Congrès International de Stratigraphie et de Géologie du Carbonifère, Madrid, 1983*, 4, 175–81.
Lonsdale, W., 1840. On the age of the limestones of south Devon. *Proceedings of the Geological Society of London* 3, 281–6.
Lowe, D.R., 1994. Abiological origin of described stromatolites older than 3.2 Ga. *Geology* 22, 387–90.
Lowrie, W. & Alvarez, W., 1981. One hundred million years of geomagnetic polar history. *Geology* 9, 392–7.
Lowrie, W. & Ogg, J.G., 1986. A magnetic polarity time scale for the Early Cretaceous and Late Jurassic. *Earth and Planetary Science Letters* 76, 341–9.
Lyell, C., 1833. *Principles of Geology, Being an Attempt to Explain the Former Changes of the Earth's Surface, by Reference to Causes now in Operation* (1st edn). vol. 3, John Murray, London, 398 pp.
McGowran, B., 1989. The later Eocene transgressions in southern Australia. *Alcheringa* 13, 45–68.
McKerrow, W.S., Lambert, R. St.J. & Cocks, L.R.M., 1985. The Ordovician, Silurian and Devonian periods. pp. 73–80 *in* N.J. Snelling (ed.), *The Chronology of the Geological Record,* Memoir 10, The Geological Society, Blackwell, Oxford.
McNamara, K.J., Rexilius, J.P., Marshall, N.G. & Henderson, R.A., 1988. The first record of a Maastrichtian ammonite from the Perth Basin, Western Australia, and its biostratigraphical significance. *Alcheringa* 12(3), 163–8.
Martinsson, A., 1977. *The Silurian–Devonian Boundary: Final Report of the Committee of the Siluro-Devonian Boundary within IUGS Commission on Stratigraphy and a State of the Art Report for Project Ecostratigraphy.* Schweizerbartsche Verlagsbuchhandlung, Stuttgart, 347 pp.
Mawson, R., Talent, J.A., Bear, V.C., Benson, D.S., Brock, G.A., Farrell, J.R., Hyland, K.A., Pyemont, B.D., Sloan, T.R., Sorentino, L., Stewart, M.I., Trotter, J.A., Wilson, G.A. & Simpson, A.G., 1989 [imprint 1988]. Conodont data in relation to resolution of stage and zonal boundaries for the Lower and Middle Devonian of Australia. pp. 485–527 *in* N.J. McMillan, A.F. Embry & D.J. Glass (eds), *Devonian of the World.* Canadian Society of Petroleum Geologists, *Memoir* 14.
Morgan, R., 1980. *Palynostratigraphy of the Australian Early and Middle Cretaceous.* Geological Survey of New South Wales, Memoir Palaeontology 18, 153 pp.
Murchison, R.I., 1841. First sketch of some of the prinicipal results of a second geological survey of Russia. *London and Edinburgh Philosophical Magazine* 19, 417–22.
Murchison, R.I., De Verneuil, E. & von Keyserling, A., 1845. *The geology of Russia in Europe and the Ural Mountains*, vol. 1, Murray, London, xxiv+8+662 pp., 7 pls.; *Géologie de la Russie d'Europe et des Montagnes d'Oural*, vol. 2, Bertrand, Paris.
Nicoll R.S., 1990. The genus *Cordylodus* and a latest Cambrian–earliest Ordovician conodont biostratigraphy. *BMR Journal of Australian Geology & Geophysics* 11, 529–58.
Nicoll, R.S., 1991. Differentiation of Late Cambrian–Early Ordovician species of *Cordylodus* (Conodonta) with biapical basal cavities. *BMR Journal of Australian Geology & Geophysics* 12, 223–44.
Odin, G.S. (ed.), 1982. *Numerical Dating in Stratigraphy Parts I and II.* John Wiley & Sons Ltd, Chichester, 1094 pp.
Odin, G.S., 1989. Ages radiométriques récemment obtenus dans la séquence stratigraphique paléogène. *Bulletin de la Société géologique de France* 8(1), 145–52.

Odin, G.S., 1992. New stratotypes for the Paleogene, the Cretaceous/Paleogene, Eocene/Oligocene and Paleogene/Neogene stage boundaries. *Neues Jahrbuch für Geologie und Paläontologie Abhandlungen* 186(1/2), 7–20.
Odin, G.S., 1994. Geological Time scale. *Compte rendus de l'Academie des Sciences, Paris,* tome 318, série II, 59–71.
Odin, G.S. & Letolle, R., 1982. The Triassic time scale in 1981. pp. 523–33 *in* G.S. Odin (ed.), *Numerical Dating in Stratigraphy Part I,* John Wiley & Sons Ltd, Chichester.
Odin, G.S. & Luterbacher, H., 1992. The age of the Paleogene stage boundaries. *Neues Jahrbuch für Geologie und Paläontologie Abhandlungen* 186(1/2), 21–48.
Odin, G.S. & Odin, C., 1990. Échelle numérique des temps géologiques, mise à jour 1990. *Géochronique* 35, 12–21.
Öpik, A.A., 1963. Early Upper Cambrian from Queensland. *Bureau of Mineral Resources, Geology & Geophysics, Bulletin* 64, 133 pp.
Öpik, A.A., 1967. The Mindyallan fauna of northwestern Queensland. *Bureau of Mineral Resources, Geology & Geophysics, Bulletin* 74: vol. 1, 404 pp.; vol. 2, 167 pp.
Öpik, A.A., 1979. Middle Cambrian agnostids: systematics and biostratigraphy. *Bureau of Mineral Resources, Geology & Geophysics, Bulletin* 172: vol. 1, 188 pp.; vol. 2, 67 pls.
Owen, M. & Wyborn, D., 1979. Geology and geochemistry of the Tantangara and Brindabella area. *Bureau of Mineral Resources, Geology & Geophysics, Bulletin* 204, 1–52.
Paproth, E., 1980. The Devonian–Carboniferous boundary. *Lethaia* 13(4), 287.
Paproth, E. Feist, R. & Flajs, G., 1991. Decision on the Devonian–Carboniferous boundary stratotype. *Episodes* 14(4), 331–6.
Perch-Nielsen, K., 1985. Mesozoic calcareous nannofossils. pp. 329–426 *in* H.M. Bolli, J.B. Saunders & K. Perch-Nielsen (eds), *Plankton Stratigraphy,* Cambridge University Press, Cambridge.
Perkins, C. & Walshe, J.L., 1993. Geochronology of the Mount Read volcanics, Tasmania, Australia. *Economic Geology* 88, 1176–97.
Plumb, K.A., 1991. New Precambrian time scale. *Episodes* 14(2), 139–40.
Preiss, W.V. (compiler), 1987. The Adelaide Geosycline — Late Proterozoic stratigraphy, sedimentation, palaeontology and tectonics. *Bulletin of the Geological Survey of South Australia* 53, 438 pp.
Riley, N.J., Claoué-Long, J., Higgins, A.C., Owens, B., Spears, A., Taylor, L. & Varker, W.J., 1995. Geochronometry and geochemistry of the European Mid-Carboniferous boundary stratotype proposal, Stonehead Beck, North Yorkshire, England. *Annales de la Société géologique de Belgique* 116, 275–89.
Roberts, J., Claoué-Long, J.C. & Jones, P.J., 1993*a*. Revised correlation Carboniferous and Early Permian units of the Southern New England Orogen, Australia. *Newsletter on Carboniferous Stratigraphy, IUGS Subcommission on Carboniferous Stratigraphy* 11, 23–6.
Roberts, J., Claoué-Long, J.C. & Jones, P.J., 1993*b*. SHRIMP zircon dating and Australian Carboniferous time. *Comptes Rendus 12 -ième Congrès International de la Stratigraphie et Géologie du Carbonifère et Permien,* 1991, *Buenos Aires,* 2, 319–38.
Roberts, J., Claoué-Long, J.C. & Jones, P.J., in press. Australian Early Carboniferous time. *In* W.A. Berggren, D.V. Kent & J. Hardenbol (eds), *Geochronology, timescales and global stratigraphic correlations: a unified temporal framework for an historical geology.* Society of Economic Paleontologists and Mineralogists Special Publication.
Roberts, J., Claoué-Long, J.C., Jones, P.J. & Foster, C.B., 1995. SHRIMP zircon age control of Gondwanan sequences in Late Carboniferous and Early Permian Australia. pp. 145–74 *in* R.E. Dunay & E.A. Hailwood (eds), *Non-biostratigraphical Dating and Correlation,* Geological Society of London Special Publication 89.

Roberts, J., Jones, P.J. & Jenkins, T.B.H., 1993*c*. Revised correlations for Carboniferous marine invertebrate zones of eastern Australia. *Alcheringa* 17(3), 353–76.
Rowen, M.K., Parrish, R.R. & Miller, D.S., 1990. The absolute age of the Eifelian Tioga Ash Bed, Pennsylvania. *Journal of Geology* 98, 282–5.
Rudwick, M.J.S., 1979. The Devonian: a System born from conflict. *Special Papers in Palaeontology* 23, 9–21.
Schimper, W.P., 1874. *Traité de Paléontologie Végétable*. 3. J.B. Bailliere, Paris, 46 pp., 110 pls.
Sedgwick, A., 1852. On the classification and nomenclature of the Lower Palaeozoic rocks of England and Wales. *Quarterly Journal of the Geological Society of London* 8, 136–68.
Sedgwick, A. & Murchison, R.I., 1835. On the Silurian and Cambrian Systems exhibiting the order in which the older sedimentary strata succeeded each other in England and Wales. *London and Edinburgh Philosophical Magazine* 7, 483–5.
Sedgwick, A. & Murchison, R.I., 1839. Classification of the older stratified rocks of Devonshire and Cornwall. *Philosophical Magazine* Series 3, 14, 241–60.
Seilacher, A., 1992. Vendobionta and Psammocorallia: lost constructions of Precambrian evolution. *Journal of the Geological Society, London*, 149(4), 607–13.
Shafik, S., 1983. Calcareous nannofossil biostratigraphy: an assessment of foraminiferal and sedimentation events in the Eocene of the Otway Basin, southeastern Australia. *BMR Journal of Geology & Geophysics* 8(1), 1–17.
Shafik, S., 1985. Calcareous nannofossils from the Toolebuc Formation, Eromanga Basin, Australia. *BMR Journal of Geology & Geophysics* 9, 171–81.
Shafik, S., 1990. *Late Cretaceous nannofossil biostratigraphy and biogeography of the Australian western margin*. Bureau of Mineral Resources, Geology & Geophysics, Report 295, 145 pp.
Shafik, S., 1994. Significance of calcareous nannofossil-bearing Jurassic and Cretaceous sediments on the Rowley Terrace, offshore North West Australia. *AGSO Journal of Geology & Geophysics* 15, 71–88.
Shergold, J.H., 1975. Late Cambrian and Early Ordovician trilobite from the Burke River Structural Belt, western Queensland, Australia. *Bureau of Mineral Resources, Geology & Geophysics, Bulletin* 153, vol. 1, 251 pp.; vol. 2, 58 pls.
Shergold, J.H., 1989. Australian Phanerozoic Timescales. 1. Cambrian Biostratigraphic Chart and Explanatory Notes. *Bureau of Mineral Resources, Geology & Geophysics, Record* 1989/31, 25 pp.
Shergold, J.H., 1993. The Iverian, a proposed Late Cambrian Stage, and its subdivision in the Burke River Structural Belt, western Queensland. *BMR Journal of Australian Geology & Geophysics* 13, 345–58.
Shergold, J.H. & Nicoll, R.S., 1992. Revised Cambrian–Ordovician boundary biostratigraphy, Black Mountain, western Queensland. pp. 81–92 *in* B.D. Webby & J.R. Laurie (eds), *Global Perspectives on Ordovician Geology*, Balkema, Rotterdam.
Snelling, N.J., 1985. *The Chronology of the Geological Record*, Memoir 10, The Geological Society, Blackwell, Oxford, 343 pp.
Strusz, D.L., 1989. Australian Phanerozoic Timescales. 3. Silurian Biostratigraphic Chart and Explanatory Notes. *Bureau of Mineral Resources, Geology & Geophysics, Record* 1989/33, 28 pp.
Swisher, C.C. III, Grajales-Nishimura, J.M., Montanari, A., Margolis, S.V., Claeys, P., Alvarez, W., Renne, P., Cedillo-Pardo, E., Maurrasse, F.J-M.R., Curtis, G.H., Smit, J. & McWilliam, M.O., 1992. Coeval $^{40}Ar/^{39}Ar$ ages of 65.0 million years ago from Chicxulub Crater melt-rock and Cretaceous–Tertiary Boundary tektites. *Science* 257, 954–8.

Talent, J.A., Berry, W.B.N. & Boucot, A.J., 1975. *Correlation of the Silurian rocks of Australia, New Zealand, and New Guinea.* Special Paper Geological Society of America 150, vi+108 pp.

Tarling, D.H. & Mitchell, J.G., 1976. Revised Cenozoic polarity time scale. *Geology* 4, 133–6.

Tozer, E.T., 1986. Definition of the Permian–Triassic (P–T) boundary: the question of the age of the *Otoceras* beds. *Memorie della Societa Geologica Italiana* 34, 291–301.

Tozer, E.T., 1994. Age and correlation of the *Otoceras* beds at the Permian–Triassic Boundary. *Albertiana* 14, 31–7.

Truswell, E.M., Chaproniere, G.C.H. & Shafik, S., 1991. Australian Phanerozoic Timescales. 10. Cainozoic Biostratigraphic Chart and Explanatory Notes. *Bureau of Mineral Resources, Geology & Geophysics, Record* 1989/40, 16 pp.

Tucker, R.D., Krough, T.E., Ross Jr., R.J. & Williams, S.H., 1990. Time-scale calibration by high-precision U-Pb zircon dating of interstratified volcanic ashes in the Ordovician and Lower Silurian stratotypes of Britain. *Earth and Planetary Science Letters* 100, 51–8.

VandenBerg, A.H.M. & Cooper, R.A., 1992. The Ordovician graptolite sequence of Australasia. *Alcheringa* 16(1), 33–86.

Webby, B.D. & Nicoll, R.S., 1989. Australian Phanerozoic Timescales. 2. Ordovician Biostratigraphic Chart and Explanatory Notes. *Bureau of Mineral Resources, Geology & Geophysics, Record* 1989/32, 42 pp.

Whittington, H.B. & Williams, A., 1964. The Ordovician Period. pp. 241–54 *in* W.B. Harland, A.G. Smith & B. Wilcock (eds), *The Phanerozoic Timescale* (A symposium dedicated to Professor Arthur Holmes). *Quarterly Journal of the Geological Society of London* 120s.

Wyborn, D., Owen, M., Compston, W. & McDougall, I., 1982. The Laidlaw Volcanics: a Late Silurian point on the geological time scale. *Earth and Planetary Science Letters* 59, 90–100.

Young, G.C., 1989. Australian Phanerozoic Timescales. 4. Devonian Biostratigraphic Chart and Explanatory Notes. *Bureau of Mineral Resources, Geology & Geophysics, Record* 1989/34, 17 pp.

Young, G, C. & Laurie, J.R. (eds), in press. *An Australian Phanerozoic Timescale*, Oxford University Press, Melbourne.

Zakharov, Yu. D., 1994. Proposals on revision of the Siberian standard for the Lower Triassic and candidate stratotype section and point for the Induan–Olenekian boundary. *Albertiana* 14, 44–51.

Zhang, Yun, 1989. Multicellular thallophytes with differentiated tissues from Late Proterozoic phosphate rocks of South China. *Lethaia* 22(2), 113–32.

Zhuravlev, A. Yu. & Gravestock, D.I., 1994. Archaeocyaths from Yorke Peninsula, South Australia and archaeocyathan Early Cambrian zonation. *Alcheringa* 18(1), 1–54.

Zijderveld, J.D.A., Hilgen, F.J., Langereis, C.G., Verhallen, P.J.J.M. & Zachariasse, W.J., 1991. Integrated magnetostratigraphy and biostratigraphy of the upper Pliocene–lower Pleistocene from the Monte Singa and Crotone areas in Calabria, Italy. *Earth and Planetary Science Letters* 107, 697–714.